iii

Sustainable Wildlife Management

Alexandra Kalandarishvili

Sustainable Wildlife Management

Myths and Misconceptions about Hunting

LAP LAMBERT Academic Publishing

Imprint
Any brand names and product names mentioned in this book are subject to trademark, brand or patent protection and are trademarks or registered trademarks of their respective holders. The use of brand names, product names, common names, trade names, product descriptions etc. even without a particular marking in this work is in no way to be construed to mean that such names may be regarded as unrestricted in respect of trademark and brand protection legislation and could thus be used by anyone.

Cover image: www.ingimage.com

Publisher:
LAP LAMBERT Academic Publishing
is a trademark of
International Book Market Service Ltd., member of OmniScriptum Publishing Group
17 Meldrum Street, Beau Bassin 71504, Mauritius

Printed at: see last page
ISBN: 978-620-0-44012-9

Contents

Abstract 2

1. Introduction 3

2. Values of Wildlife 4

3. Conservation VS Preservation 10

4. Unsustainable use of wildlife: Poaching and the Illegal Trade in Wildlife 14

5. Sustainable Consumptive Use of Wildlife: Characteristics of Responsible Hunting 23

6. Biological and Ecological Benefits of Hunting 25

7. Predator Management 28

8. Case Study: Methods of Feral and Stray Cat Population Control: Trap Neuter Return vs. Trap and
Remove 34

9. Human-Wildlife Conflict 41

10. Driving the economy 43

11. Game Meat vs Produced Meat 45

History of hunting 45

Food security 46

Nutritional value 48

Ethics 49

12. Conclusion 51

References 52

Abstract

Hunting is often viewed as a cruel and unethical activity. True motives behind hunters' actions are overlooked and most of the times misunderstood. All over the world ban on trophy hunting is a common demand, however the consequences of the ban are highly underestimated. Unmanaged and unregulated wild animal populations are the main reason for human-wildlife conflict in both rural and urban areas. Hunting is a vital tool for conservation, and when done properly it not only resolves conflicts arising between the wildlife and the human worlds, but also plays a key role in recovering endangered species, and ends up greatly contributing to nations' economies. Hunting is a method by which natural populations are kept at numbers that the available habitat can support. For species that frequently cause damage to private property, hunting helps maintain populations at levels that are compatible with human activity and land use. It is important to consider wildlife as a renewable natural resource with a surplus, which is harvested by hunters seasonally. Quotas and specific hunting seasons for each species ensure that the population is not overharvested. It is important to point out that revenue generated from application fees, hunting permits, and stamp sales have helped many wildlife species and habitats to recover from devastating conditions. In 2008 hunting was estimated to be worth €16 billion in Europe alone, which directly financed habitat improvement and maintenance efforts, research, public information/education, and wildlife law enforcement works. Moreover, sustainable hunting has significant cultural value in many societies, it employs local resident in numerous activities and involves them in decision making processes.

Keywords: Conservation, Hunting, Management, Sustainability, Wildlife,

1. Introduction

The era when technological advancement has reached its peak, and in the time when urbanization has become unstoppable, natural habitats and its inhabitants are facing a great risk. Across the continents declining trends in the most treasured species are no longer unusual. Disappearing biodiversity is resulting in the collapse of natural order and thus, ecological services so far providing necessary means of survival will soon no longer stand available. Defining a solution to prevent the biological diversity form perishing has never been so urgent.

Wildlife management is a key part of biodiversity conservation. However, its methods remain to be somewhat controversial among the public. On one hand, hunting – being a key component of wildlife management – is endorsed by many rural communities. Its contributions to the wellbeing of the society are obvious, as hunting provides high quality food markets, regulates the problematic wild animal species, improves biotopes and enriches the economy of the nations. Hunting not only reduces damage brought to the agricultural and forest lands by wild animals, but also plays a major role in recovering and preserving threatened fauna species.

However, trophy hunting is met with an increasing opposition from animal rights and welfare organizations, which often results in strong clash of ideals. False assumptions around the issue of trophy hunting are common, consequently this activity is frequently misunderstood. Animal rights activists often fail to understand the true motives of hunters and are demanding a ban on trophy hunting. In some cases like Kenya and India for example, hunting is either completely banned or significantly limited. Moreover, these organizations are making an attempt to influence policy makers and the media across the globe, however, the images presented of trophy hunting are often inaccurate and most of the time deliberately misleading. Furthermore, these organizations have strong public support and often times, sound, scientific facts are over-shadowed by pure emotion and personal feelings towards individual animals.

3

Despite their commitment to species and conservation, and despite many success stories hunters have told, hunting remains to be an activity widely and frequently condemned by the general public. There are many misconceptions and myths people believe when it comes to the topic of trophy hunting. It remains of utmost importance to outline and analyse the common misconceptions of hunting, and point out the very active role this activity plays in endangered species recovery programs, and in resolving the human-wildlife conflict. Only a small portion of the public recognizes the contributions trophy hunting is making to the economy of developing countries, and very few are aware that hunting for many regions is a key element for local culture and traditions determining the lifestyle of indigenous communities.

However, despite the many different views and mindsets, one thing is clear, that wildlife is loved and valued almost instinctively by everyone. Ensuring its wellbeing is a fight everyone is willing to fight. Such an attitude raises an interesting question – why? What is it about wildlife, that the thought of losing it is so unbearable for humanity? Why do we go out of our way to protect it? What are its benefits, and why have we come to value wildlife so much?

2. Values of Wildlife

The fundamental question in wildlife management is: What is the ultimate goal of wildlife management? why do we try so desperately to manage wildlife? We put so much time and money in wildlife management that there has to be something truly important we are trying to accomplish. Between the 2001 and 2008 total annual expenditure on global biodiversity conservation amounted to $21.5 billion (Waldron et al. 2013). In the United States, between 1992 and 2001, the federal government spent about $22.6 billion on land conservation. $15 billion was invested by The Conservation Reserve Program (CRP) over the same 10-year period. Among the permanent protection programs, the Land and Water Conservation Fund was the top spender for land acquisition which contributed about $1.6 billion. The Wetlands Reserve Program

4

invested the most toward conservation easements amounting to $844 million. In addition, States spent close to $9.5 billion on land conservation between 1992 and 2001 (Lerner et al. 2017). While these are not very recent figures, they clearly show the extent to which people care for the conservation of land. After this, it is very difficult to imagine wildlife management without some kind of goal.

There are couple of conclusions we can draw from such large scale actions for wildlife and habitat management. 1) We as a society have some sort of vision of what is best for wildlife, and are working towards fulfilling that vision. 2) We believe that this vision will not be fulfilled without our actions. 3) We are certain that the vision we have will be achieved only IF we intervene with the natural order. This, in turn, justifies many human interferences and disturbances caused to the natural world.

So what is our vision, our goal when we manage wildlife? To answer the question at hand, it is important to understand that wildlife has values. Many values, that influence our perception and attitude towards them. It has *financial value* that comes from activities like tourism and trophy hunting. In 2016, expenditures by sportsmen and women amounted to $81 billion, while wildlife watchers spent $75.9 billion in the United States. Watching, feeding, and/or photographing wildlife was enjoyed by 86 million people that same year (U.S. Department of the interior, 2016). Wildlife thus, has the *aesthetic and therapeutic values* – inspiring and enriching our day to day lives.

It has *subsistence value* where animals are raised and/or hunted for food and fur. This was actually the main use of wildlife before the development of the agriculture. Even today, people globally satisfy some part of their subsistence needs by harvesting wildlife.

Animals have been of high importance for scientific community as well. Wildlife provide clues about our past and our future. Most of what we know about the natural world is the result of observing and studying species and their behavior. Some of the

5

species are used as indicators, to monitor the environmental health. As an example, spotted owls are used as indicators to see whether or not there are enough old-growth forest in the Pacific Northwest of the United States. Spotted owls have a relatively big home range and would be the first one whose populations would suffer in case the old-growth forests become scarce. Therefore, if there are enough old-growth forest to harbour a healthy population of spotted owls, it is of sufficient amount to meet the needs of other species as well (Conover, 2001.).

Wildlife has a strong *cultural value,* as many great cultures regard animals as powerful symbols. One must only look at flags and coat of arms to realize the importance animals hold in nations' pride. Wildlife also has what is called an *existence value*, which refers to its potential to become valuable at some point during the future. Even the most currently useless lands are exchanged for money because of their potential usefulness. Similarly, wildlife may have values in the future which are unrecognized today (Conover 2001).

Interestingly, there is a connection between the *Historic value* of wildlife and the mankind's fear of change. We are not fond of things changing around, we experience a type of uneasiness when we think how human actions might change the environment. We are almost reassured of our well-being when there is minimal change in the natural environment. This is also the case with wildlife. Humanity feels at peace knowing there are still wild areas with thriving wildlife populations, just like there has been for thousands of years. People felt at ease, proud even, like the world was made right somehow when for example, wolves were reintroduced in the Yellowstone National Park in 1995 (Conover 2001).

Sometimes the *historic value* and the *existence value* are assumed to mean the same, but there is a large difference between the two. Conover 2001 gives one very interesting example which relates back to the reintroduction of wolves in the Yellowstone National Park. The question is asked – why was it specifically the *wolf* that was reintroduced in the Park? The reason was clearly not to preserve the wolf species, as other populations were found elsewhere at that time, and the wolf was not considered a threatened species. If the goal is to increase existence value of wildlife to the maximum, then perhaps the best way would have been to release a predator which is far more endangered and is threatened with extinction. The argument goes on to give an example of the Siberian tiger which is far more endangered than the wolf and could also, very well survive in the Yellowstone ecosystem. However, the idea of having tigers in the United States is rather strange, even worrying as historically, they were never encountered there. Presence of tigers in the Yellowstone would surely increase the existence value, although dramatically decrease the historic value. In this example maintaining the historic value would be more important than increasing the existence value (Conover 2001).

Perhaps, the greatest value that wildlife has is the fact that it is a key part of our global biodiversity which sustains life – a complex web of living organisms that allows each life form to survive and reproduce. It is a frequently asked question, what would happen in biodiversity disappeared? What would happen to humanity? It is clear by now that our existence depends on many factors, and one of the most obvious and important factor is the environment. The slight reduction, not even the loss, but the slight reduction of biodiversity tremendously impacts our livelihoods, our health, the economies and our lifestyle. One must not go too far but remember the goods and services the ecosystems provide – clean air, fresh water, food, raw materials, genetic resources, energy, medical resources – the list goes on and on.

Wildlife, while bringing obvious benefits, also has what can be considered "negative values". The *negative value* would include, wildlife attacks, vehicle collisions, crop and forest damage, zoonotic disease transmissions, private property damage and so on. In order to determine the true value of wildlife all that it takes is adding up all the *positive values* and subtracting all the *negative values*. This will determine the *net value* of wildlife.

The answer to the question at hand – what is our goal when we manage wildlife – it is to increase the net value of wildlife for the society. Almost all efforts in wildlife management try to achieve this by increasing the positive values that the wildlife has. Managers try to increase the value of game species by managing them to produce maximum sustainable yield. In this case, hunters for example, will be able to attain the maximum recreational and subsistence value. Additionally, the *wildlife damage management* aims to increase wildlife's net value by reducing the negative values of wildlife. Therefore, wildlife management can be considered a science and a practice which aims to enhance the value of wildlife by decreasing its negative values, and increasing its positive values (Conover 2001).

One of the debatable topics is whether the wildlife ought to be managed for the benefit of wildlife itself or for the benefit of humanity. It is often that many environmental ethicists consider wildlife as being of the *intrinsic value* which is valued for its own sake or simply for existing. In his Land Ethic, Aldo Leopold argues that humans should not view themselves as "conquerors" of land but rather as a plain member of biotic communities. Leopold's philosophy talks about extending ethical consideration to other ecological components such as soils, water, plants and animals. What this means is that humanity should selflessly strive to preserve the stability, integrity and beauty of the biotic community. Charles Krauthammer held a different position and argued for "a sane environmentalism".

> "How are we to choose among the dozens of conflicting proposals, restrictions, projects, regulations, and laws advanced in the name of the environment? Clearly not everything with an environmental claim is worth doing. How to choose?" – Saving Nature, but Only for Man, 1991.

Krauthammer argued that to make a reasonable choice it is essential to divide the environment into two categories; environmental luxuries and environmental necessities.

Luxuries being those things that would be nice to have if it does not come at a price. Necessities, being the top priority, are things that are absolutely crucial for human well-being. The idea of sane environmentalism binds humans to preserve nature solely for their own benefit.

Whichever one aligns with personal morals is a question of individuality, however, there are two crucial principles to take into consideration when discussing wildlife management philosophy. First is that **wildlife (as any other natural resource) is not inherited from our ancestors but rather borrowed from our descendants.** What this means is that no generation, ours included, has the right to take such action that will diminish the value of wildlife resource for generations to come. Thus, the second principle of wildlife management is that wildlife management should be used as a tool with which to increase the value of wildlife.

This is not something that is easy to foresee, it is very hard for each generation to know what component of the natural resource will be important for the future generation. The way wildlife is viewed also changes from generation to generation. Former generations valued those wildlife species that had high economic value like bison, elk, deer, and beaver in the United States. Today species are given value based not only on economic value but also on the role they play in the ecosystem and because of the aesthetic value that they have. So how do we foresee the interest of future generations? And is it actually possible? The answer is no, it is not. Therefore the actions we take must be well calculated so that they do not cause irreversible damage.

Upon realizing the criticality and the complexity of the situation, two separate ideas emerged, each of them giving a different formula for safeguarding the wellbeing of the ecosystems. These two ideas, or practices are known as *Conservation* and *Preservation*.

3. Conservation VS Preservation

Aldo Leopold, often referred to as the father of ecology, and the one who was a significant figure in wildlife management, called for wilderness protection and land ethic. Another significant environmental science icon, Rachel Carson's Silent Spring greatly inspired modern environmental movement. Their works touched the hearts of thousands, making them realize the importance and necessity of the natural world. It is difficult to find a person who does not care for wildlife, and does not want to see it flourish. But there are different point of views by which means should the biodiversity thrive.

Conservation and **preservation,** two separate ideologies strive to safeguard the natural world, the two share a vision of healthy ecosystems with high biological diversity. Most of the times, terms *conservation* and *preservation* are even used interchangeably and are assigned the same definition, however, there is a big difference making the two radically different concepts.

To put it simply, conservation follows the idea of responsible management of natural resources for human benefit, however, the knowledge that these resources are vulnerable and exhaustible is always kept in mind – it seeks the sustainable use of natural resources. The practice of *conservation* tries to fulfil the biological, recreational, cultural and economic needs of the people, as well as make sure that wildlife stays safe of any potential threats. It does not oppose the harnessing of nature for human progression, provided it be carried out in a sustainable way. Conservation views economy as a crucial factor and adapts policies in ways that are beneficial also to the economy.

Preservation stands on the opposite side as it seeks to eliminate human impact on nature altogether. Upon researching preservation, it is likely to come across the idea of *deep ecology* – a philosophy which views all species as having equal worth and right to natural resources. Given the condition of the Earth today, it was only natural for such

10

strict preservative views to have emerge. Exploiting natural resources and utilizing them without set limits for housing, farming and tourism led to environmental catastrophes we witness today.

While fundamentally different at its core, we shall not choose between the two, in fact, both preservation and conservation play a crucial role in biodiversity security. The problem arises when some of the conservation and management methods are popularized as cruel and unethical, and given the bad reputation is does not deserve. The famous idea of *"leave the nature alone"* or *"nature knows best"* gets in the way of that very *sustainable development* that radical nature protectionists are striving to achieve.

Many of the organizations that claim to be protectors of rights of animals are calling for hunting ban. The influence they have over the general public and over policy makers is quite powerful, which is worrying for the scientific and hunting communities. Removing hunters from conservation will result in massive cut in both budget and human resources thus leaving wildlife population unprepared for any potential threats.

Nevertheless, it is an interesting scenario to explore – what would happen if hunting were to be banned all together? For good.

For starters, hunting is one of the main conservation tools which is supported by over 7 million hunters in Europe (FACE. 2010) and 11 million (US fish and Wildlife Service. 2017) in the United States. Once the total ban is placed on hunting, not only will it result in a big dissatisfaction within the hunting community, but now the many organizations that primarily focus on wildlife and habitat conservation – like Ducks Unlimited, Rocky Mountain Elk Foundation, Pheasants Forever, Quality Deer Management Association, The Ruffed Grouse Society and many other alike – will no longer have their key supporters.

11

For more than 50 years, The Ruffed Grouse Society was an active participant in the restoration and improvement of 500,000 acres of land on federal, state and country lands[1]. At the same time, Pheasants Forever takes part in land acquisitions in order to permanently protect critical habitat for upland wildlife, while simultaneously opening the areas to public recreation. As a result of cooperating with local, state and/or federal natural resource agencies, more than 187,000 acres are being protected and open to the public because of Pheasants Forever's participation[2]. Ducks Unlimited managed to restore hydrology to eight wetlands, which were drained decades ago on a 296-acre parcel of native grassland and wetland habitat, near the southern shore of Waubay Lake. About 39 wetland acres were successfully restored by plugging drainage ditches, bringing back their natural hydrology[3]. These types of organizations are incredibly important as they raise millions of dollars to support wildlife habitats. In Europe alone trophy hunting is worth more than €16 billion (Middleton 2014). Take that away and you are left with very little to no financial means with which to support species and manage their habitats.

With habitats degrading, wildlife will move elsewhere in search of food, and that raises concern. It is worth noting, that the home range size of an individual animal strongly depends on the habitat quality. Individual's home range is smaller in a high quality habitats, as all the essential resources (food, water, shelter etc.) are abundant and found nearly everywhere. In such habitats individuals are not required to travel long distances in search of these resources. On the contrary, in a low quality habitat the resources are scarce and scattered, thus individuals tend to travel longer distances and more frequently. This is very worrying because in their search for food, some of individuals are sure to move towards human settlements. Naturally, moving close to the settlements

[1] https://ruffedgrousesociety.org/habitat/
[2] https://www.pheasantsforever.org/Habitat/Why-Habitat.aspx
[3] https://www.ducks.org/conservation/gpr/sd-wetland-restoration-keeps-lakes-clean?poe=conservation

fuels the human-wildlife conflict in form of property and crop damage, as well as wildlife attacks and vehicle collisions. The dissatisfaction in the hunters' community is expanded to dissatisfaction in other civilians and farmers.

Now, with their top predator gone, some of these wild populations might thrive and increase in size, but thriving populations are the ones that suffer the most, as they exceed the biological carrying capacity and become overpopulated. With an increasing number of individuals, there is higher competition for food and other resources resulting in mass death of those populations due to starvation. On top of that, because of high animal concentration on a small area, diseases will be spreading like wildfire among individuals killing wildlife one by one.

Wildlife slowly destroying itself is an unbearable thought. However, wildlife is not the only one that will suffer, humans will too. It is important to understand that in conservation the wellbeing of the society is just as important as the wellbeing of wildlife. Many rural communities depend to some extent on wildlife for subsistence.

Nevertheless, the idea that humans should back off and let nature run its course is still an actively used argument by people who favour preservation over conservation. After all, the ecosystems – they say – were doing perfectly fine until humans interfered with natural order. This argument is not too far away from the truth. Ecosystems were indeed fine *until* the constructions of houses, factories and roads. Nature was fine *until* intensive agriculture came into the picture with the use of pesticides, insecticides and herbicides. Nature was perfectly fine *until* the emissions of greenhouse gasses that lead to events like climate change.

It comes as an unfortunate reality that in this heavily human-modified world it is especially hard to reach and maintain a balance between us and wildlife. If species are to be enjoyed for a long time, humans must intervene and manage nature itself, modify habitats, control the predator populations and regulate the size of ungulate populations. Leaving nature to run its course would not only be a mistake, but will be an irresponsible way to deal with environmental issues on our part. Leave the nature to run its course and the chain of events that will follow will endanger the wild world together with the human world.

4. Unsustainable use of wildlife: Poaching and the Illegal Trade in Wildlife.

Explaining the difference between hunting and poaching is what most hunters had to do at some point. It is often that certain terms and phrases are misinterpreted and used incorrectly. Though at times it might feel frustrating, it is crucial that this distinction is clearly made.

Hunting is the *act* of setting out into the wild in hopes of attaining a kill. However, the act itself is not what matters in the world of conservation. What matters is the *purpose* behind setting out into the wild. There are many reasons why one might set out hunting, and these reasons are what lead wildlife to different circumstances. Poaching, culling, trophy hunting and subsistence hunting – all involve the *act* of killing an animal – all of these activities, involve the *act* of hunting but each of them is taking wildlife down a different path. Most of the time poaching and legal hunting are assumed to be the same and the negative label is attached to both of them popularizing both as a same negative activity.

Not only is poaching a crime recognized by law in many countries, it is unsustainable exploitation of wildlife that decimates wildlife populations. It is the complete opposite of recreational hunting which aims to maintain population abundance, distribution, structure and behaviour of species. Trophy hunting aims to maintain genetic diversity – by encouraging maintenance of subpopulations – and improve conservation status of endangered and threatened animal species (IUCN ESUSG 2006).

Poachers are assumed to be selfish, evil, and corrupt individuals – people who are indifferent to wildlife, people who do not think twice before causing an irreversible damage, people who would do everything, anything for self-gain. While this statement may be true to some extent, it is not necessarily the whole truth. A number of rural communities, rely heavily on wildlife for subsistence. Most of the times, Africans are forced to poach due to their social insecurity (Mahoney 2019). However, few animal rightists regard this as an adequate excuse for hunting, and subject African poachers to their harshest criticism.

In Africa today there are two types of bushmeat poachers (African forests and savannahs are often referred to as bush, hence the term Bushmeat). On one hand there is the traditional hunter, whose primary goal is to feed himself and bring food to his family. These poachers are settled in highly rural areas and live in the type of areas most people would describe as extreme poverty. This hunter uses simple tools and methods for killing, and his interaction with wildlife is very similar to those of our early ancestors. Most of his family's protein comes from bushmeat, but it is hardly ever in sufficient amounts (Mahoney 2019). As much as 30% to 85% of daily protein intake of Africans come from bushmeat. In certain countries like Gabon, Guinea and Benin where domestic meat is expensive and limited, wild meat is a great substitute which is cheap and locally available (Williamson and Lonneke Bakker). While occasionally trading his prey for other goods and types of food, his main focus while hunting is subsistence, not profit. Although a poacher, his motivations are universally understood and his actions are typically judged less harshly (Mahoney 2019).

On the other hand, there is a commercial bushmeat hunter. Bushmeat today is a luxury food consumed by people that are considered elites. With its demand on the rise bushmeat trade has become a highly profitable business. Commercial bushmeat hunting forms the basis of a multi-billion dollar international trade, and the markets are not African alone. Species of wide variety are being imported from Africa to small countries like Switzerland. About 1,013 tons of illegal meat, and 8.6 tons of bushmeat is being imported in Switzerland annually. Between 2008 and 2011, 5.5 tons of illegal bushmeat was seized in the Swiss international airports, out of which 1.4% was bushmeat (Falk 2013). A study which searched the passengers' luggage in Roissy-Charles de Gaulle airport, Paris, France for illegally imported bushmeat, concluded that approximately 5 tons of bushmeat in smuggled in Europe via Paris. This amounts to about 260 tons of illegal bushmeat being smuggled in Europe every year (Chaber 2010).

It is clear, that for a commercial hunter, bushmeat is his livelihood and his main goal, unlike the traditional hunter's, is to make *profit*. While some of his hunted animals will be used for subsistence, he mainly provides for his family with money earned through organized, and clearly unregulated illegal trade. These poachers often join migrations when animals are traveling long distances, and leave the protected areas and happen to move closer to the community lands. Unlike the traditional hunter, this poacher kills in very high numbers using more advanced tools and methods. The stock piles up, consisting of hundreds of animal carcasses. The commercial bushmeat poacher is paid after the employers have picked up their products and transported them beyond country borders (Mahoney, 2019).

The list of animals affected by poaching and illegal trade is a long one, including gorillas, chimpanzees, elephants, antelopes, crocodiles, fruit bats, porcupines, pangolin and many other species coming from different taxa. These animals are killed with inhumane and illegal methods, some of which are non-selective – wire snares, firearms, and poison, are some of the methods used by poachers. The growing problem of the illegal, unsustainable hunting for bushmeat is worsened by growing urbanization, namely by the construction of roads to enable logging and mining operations. Such "human-friendly" environment allows poachers easy access to remote parts of the forests[4]. Unlike poachers, legal hunters avoid human-built roads and easy routes. They embark on an adventure which does not ensure an easy taking of an animal. Hunters normally head towards remote and uninhabited areas, which are otherwise undisturbed.

Despite the fact that this practice of illegal killing and illegal selling of bushmeat has the benefit of earning quick cash, it is ultimately unsustainable, and denies opportunities for the wider community and future generations to benefit from wildlife. The U.S Fish and Wildlife Service states that the most immediate threat to the future of wildlife in Africa and around the globe is not habitat loss, but rather the illegal trade in, and the consumption of bushmeat[4]. It is estimated that about quarter of all mammal species which are faced with extinction are killed and sold for bushmeat (Mahoney 2019). Thousands of kilograms of primate parts, antelope, and other bushmeat are illegally brought in United States and Europe every year for different cultural events[4].

This practice is not only harmful for the wildlife but it poses great risk for humans as well. Bushmeat has been linked to numerous infectious disease outbreaks, including Ebola, HIV, and monkeypox. The diseases are transmitted to humans through hunting, butchering and consumption of contaminated bushmeat. While consumption of wild African meat is mentioned most frequently, the highest risk of disease transmission is associated with exposure to body fluids during handling of the meat. The highest transmission risk occurs during the butchering process, as it involves working with sharp tools which leads to cuts during the process (Kurpiers. 2016).

Poaching for bushmeat has always been a great problem. While bushmeat crisis is a growing concern in east and Southern parts of Africa, West and Central Africa experience the worst of it. Unsustainable hunting for bushmeat is considered a single most important immediate threat to wildlife, considering the fact that approximately one million metric tons of bushmeat is consumed in Central Africa each year[4]. The *Empty Forest Syndrome*, it is a term used to describe African forest that remain to be

---

[4] https://www.fws.gov/international/wildlife-without-borders/global-program/bushmeat.html

structurally intact but are devoid of wildlife as a result of illegal bushmeat trade. The illegal wildlife activities do not only affect species that are directly poached, but it is having an effect on the whole of environment. The loss of wildlife biodiversity disrupts natural ecological processes such as seed dispersion and pollination which in turn leads to even greater ecosystem degradation.

Since 2002, when the international monitors began to keep detailed records, it was concluded that African elephant poaching reached the highest levels between 2011 and 2014. 100 000 elephants were killed illegally for their ivory across the African continent between 2010 and 2012, as a result the elephant numbers were reduced to 420 000 individuals in the whole of Africa. However, despite some success with reducing elephant poaching since 2011, in some African areas illegal killings continue at such high rates which exceed the natural reproduction ability of elephants[5].

Apart from the direct threat of poaching, growing human population has a devastating effect on the African wildlife. The wild habitats are shrinking and disappearing at an alarming rates, due to forests and savannahs being cleared for infrastructure development, agriculture, and extractive industries, such as logging and mining. As a result of habitat loss and food scarcity, hungry animals like elephants move closer to the settled areas in search of food and water. This in turn results in an intense human-wildlife conflict. With little left to do, in efforts to save their private property, harvest, and their families, locals often take extreme measures to get rid of the animal that endangers their livelihood.

Though a serious matter, poaching is only a beginning of a more serious problem that comes in form of illegal trade in wildlife and its products. This multi-billion dollar business of global wildlife trade is not only a big threat to biodiversity, but also to the economy. However, little is known about the scale of the illegal wildlife trade. Ideas and hypothesis have been put forward, which estimates this trade to be between $5 and $20

---

[5] https://www.fws.gov/international/wildlife-without-borders/african-elephant-conservation-fund.html

billion a year. This places illegal wildlife trade among world's largest illegitimate businesses, next to narcotics (Wyler and Sheikh, 2008).

Convention on International Trade in Endangered Species of Wild Fauna and Flora (CITES) is an international convention that takes the responsibility upon itself to regulate this illegal trade. Although some of its methods are subject to hot debates, the convention comprises of 150 parties that regulate the trade of more than 30 000 endangered plant and animal species. The main goal of CITES is to ensure that wildlife species are not endangered due to international trade. CITES regulates the trade by listing species in one of its three appendixes. Appendix I prohibits any kind of international trade in its listed species – most of these species are recognized to be threatened with extinction. Species that are listed on the appendix II are not under the immediate risk of extinction, however unless the trade is controlled, the risk of them becoming threatened is high – species listed on the second appendix require both the importing and exporting permits from countries, which proves that the specimen were acquired under legal circumstances. Finally, Appendix III contains species that are protected in at least one country which has asked other CITES Parties for assistance in controlling the trade.

Unlike Appendix II and III which are less concerning, Appendix I causes great worry among conservationists. It is argued that the "up listing" of a species to Appendix I status can potentially increase the illegal trade activity, which is the very thing the convention is fighting against. Closing down the market will only make these products more valuable because of its rarity and make them more appealing to the buyer – in other words the demand on these products will increase if not remain the same. This in turn will intensify poaching in order to satisfy the growing demand. This has been the case with the rhino horn and elephant ivory trade.

South Africa has the world's largest white rhino population, and hosts one third of the black rhino. However, the country also has an extremely high rate of poaching. It was

19

estimated that about 260 rhinos were poached in 2010. This translates into approximately one rhino being illegally killed every day. Such high rate greatly worries conservationists and African rural communities who depend on local wildlife for subsistence (Waddell, 2010). Since 1976, the trade in rhino horn has been banned in hopes to reduce the illegal killings and selling of rhinos and its products. However, the result was quite the opposite. Since the ban was issued, all five species of rhino have been listed as either endangered or threatened with extinction. Some groups associate this rapid drop in rhino numbers with the trade ban (Abensperg-Traun, 2009). The loss of this species will not only devastate the ecosystems and local communities but it will also reflect on our ability or inability to address critical environmental issues. Conservationists are worried that the current laws and regulations related to the international trade of rhino horns does more harm than good – that it encourages criminals to find ways of bypassing these regulations (Larson, 2010). In 1987, CITES Parties took it a step further and extended the ban by also outlawing the domestic trade in rhino horn, however, the ban failed to include the stockpiled horns, which turned out to be a big mistake. As a result, because of the difficulty to distinguish between the stockpiled and newly imported horns, the trade in rhino horns continued on more intensely.

When it comes to rhino situation, many argue that banning the trade and pushing for the *protectionist* ideology is not the most effective way to solve the problem of poaching. The uplisting, political instability, corruption, and lack of political will and resources to control poaching are primary reasons why rhino populations are still on the decline today (Abensperg-Traun, 2009). The declining rhino populations is also related to increased prices for its horn on the black market (Leader Williams, 2005). Within only two years of uplisting the species, the horn price increased by more than 400% on the Korean markets which in turn resulted in sharp increase in poaching rate and illegal trade in black rhinos (Rivalan et al. 2007).

China has been the world's largest importer and manufacturer of rhino horn and its products, yet it has also been an active member of CITES since 1981. As far as the rhinos are concerned, there is a weak evidence that Appendix I listing has the ability to improve the conservation status of the species because even today, despite the ban, trade continues (Abensperg-Traun, 2009). In addition the difficulty to track the products in the illegal trade makes the problem worse.

Another actively discussed species at the CITES Conference of Parties (CoP) are the African elephants. Current appendix I listing of elephants is also a topic of continued debate among conservationists and protectionists.

In 1980s the elephant population in Zimbabwe skyrocketed, and reached extremely high numbers. The damage brought by such high elephant numbers was clearly visible – more and more trees were found knocked over, farmers were complaining more frequently about their damaged crops and property, and woodlands were rapidly transforming into grasslands. Elephants were destroying their own habitats, and were now dying as a result of starvation. This also resulted in waste of meat in already protein deficient country. The situation also became the matter of ethics, in allowing the animals to die. Since water in the Hwange National Park as provided artificially and was not a natural environment, it was absolutely crucial for the area to be managed (Mackintosh 2015).

Zimbabwe's National Parks and wildlife management authority has the responsibility to protect all ecosystems and its components, without favoring individual species. Damaged woodlands also impacted the wellbeing species other than the elephant. Such critical circumstance that was evident in Hwange National Park led the authorities to develop a culling program, in order to reduce elephant populations and allow the forests to recover. For culling to be effective, more animals should be removed than are recruited in the population. There was a very little risk in overculling, as it takes about

15 years for elephant populations to double while it takes 10 times more for the woodlands to recover (Mackintosh 2015).

In 1983 the goal was to reduce the number of elephants from 23 000 to 12,000 – the level at which a serious damage in the park first caused concern. However, it soon became evident that even 12 000 individuals are too many for the state of condition of the environment. Culling is itself an expensive operation (culling around 1000 elephants is in the region of around $300 000), and at the time ivory trade paid for such management. However, in 1989 the elephant was put on CITES Appendix I resulting in a sudden cut in funding. No funding meant insufficient management, and no management meant increasing elephant populations which resulted in intense human-wildlife conflict among other problems like habitat degradation. While some money can be recovered for the local sales of ivory, meat and skin, the only way to recuperate the expense of culling operation is either through international sales of ivory or foreign funding (Mackintosh 2015).

Two one-off sales of raw ivory have taken place since elephant was listed on appendix I in 1989. The first was in 1999 when Botswana, Namibia and Zimbabwe sold ivory to a single buyer – Japan. The total amount of funds received from the auctions was approximately $5 million. The second took place in 2008 when Botswana, South Africa and Namibia and Zimbabwe sold to China and Japan. The total amount of funds received from this auctions comprised of about $15.5 million. Funds that would be generated through these sales were officially planned to go to elephant conservation programs (Mackintosh 2015).

It is wrong to assume that banning hunting, banning the trade, and stopping using wildlife resources will bear any positive results. Responsible, sustainable utilization of all natural resources – not just wildlife – is what safeguards our livelihood as well as the natural world. In reality, sustainable hunting has been playing an important role in recovering endangered species and keeping populations stable and fit. For some, it surely does sound absurd that hunting works in favor of endangered animals.

Hunting is indeed a complex topic and it is often difficult to draw the line between different types of hunting – where does subsistence hunting end and where does trophy hunting begin? Sustainable hunting and poaching are topics of discussion that have no end. Is the unregulated hunting that falls within the legal framework in many countries, but which is definitely unsustainable, is that considered poaching? Should we be defending the type of hunting which uses non-selective methods but which have been part of hundreds of thousand year old cultural traditions? And what if the type of overhunting contributes to the food security of entire communities. Is that also defensible? Many questions are difficult to answer, and that fact leaves people disappointed and wondering.

5. Sustainable Consumptive Use of Wildlife: Characteristics of Responsible Hunting

Hunting is one of the key components of wildlife management and has many positive effects on ecosystems when done properly. Despite the opposition, hunting is playing an important role in sustainable development. The idea of sustainable hunting is to maintain animal populations at an economically and ecologically healthy level for the amount of space that is available. In order to accomplish optimal population size and avoid posing risk to the species, authorities encourage *quotas* and *hunting seasons* for specific species as a form of regulation which limits daily and seasonal harvest. In addition, hunters are committed to the principles of *fair chase* and willingly restrict their hunts to individuals of a certain sex and age group (Conservation visions 2017).

One of the common arguments against hunting is the idea of overhunting. This is a real issue, as overhunting poses a very real and a serious threat to wild populations. Being aware of this danger, wildlife professionals have encouraged *bag records*. Bag records are crucial for understanding population dynamics and play an important role in adapting appropriate management plan. Trophy hunting is based on the idea of *adaptive management* which continuously monitors and evaluates the current situation and if necessary, revises the management plan completely (IUCN ESUSG 2006).

Keeping populations fit is one of the priorities of sustainable hunting. Food scarcity has devastating effects on wildlife communities by weakening animals' immune systems and making them prone to diseases. Diseases rapidly spread as a result of predation, reproduction and by direct contact. It is quite frequent that viruses like African swine fever virus is transmitted from wild species to domestic ones – decimating domestic animal populations. In order to avoid transmission of the disease within and between populations, it is necessary to remove the weak and sick individuals giving reproduction opportunity to stronger ones. This system keeps the populations fit and at the same time reduces the chance of wild animals causing damage to the rural communities.

In most cases wildlife species are illegally killed if they provide little benefit to local communities or cause them substantial damage. The individuals are sold illegally as food, or as commercially valuable products while their habitats are degraded and used for different purposes (IUCN SSC 2012). Trophy hunting has the ability to address this issue by making wildlife of higher value to the local people. Sustainable hunting views socio-economic sector as one of its main considerations. In this context hunting aims to maintain the abundance and distribution of hunted species to the level that is compatible with the interests of socio-economic sector. The system also encourages local employment and participation of local hunters. It also supports optimizing utilization of game meat and other products. At the same time sustainable way of hunting ensures that the cultural, historical and artistic values of hunting and wildlife are well preserved (IUCN ESUSG 2006). It is important to note that trophy hunting involves taking small numbers of individual animals and does not require highly developed infrastructure. The hunts are therefore high in value but low in impact compared to other types of land uses like agriculture or tourism (IUCN SSC 2012).

24

6. Biological and Ecological Benefits of Hunting

The world where humans have taken the lead becomes an unfortunate reality for the most unique species that are tittering on the brink of extinction. Defining a viable solution to avoid mass extinction has never been so urgent. Sustainable hunting has been playing an important role in keeping populations stable and healthy, while also recovering species that are facing the threat of extinction. In this extremely urbanized and modernized reality, finding a balance between us and wildlife has become exceptionally challenging.

North America's wealth in renewable natural resources is what led to its exploitation by early European settlers. Its abundance in resources allowed the expansion and settlement of its people across the continent. In 1820, 5% of the Americans lived in cities; 40 years later, in 1860 the number has grown to 20%. This 4-fold increase marks the greatest demographic shift that has ever occurred in America. Wildlife markets arose to satisfy subsistence needs of the growing urban population (Organ et al 2012). Game meat quickly became prestigious food for the elites, while fur and feathers became fashionable accessories.

It comes as no surprise that such rapid modernization of the land significantly impacted wildlife and their habitats. Before the arrival of the Europeans, elk were the most widely distributed species whose geographical range extended from southern Canada to Northern Mexico. However, by the mid-1800s elk started to decline in the eastern United States, and soon all over the country. An estimated 10 000 000 elk rapidly dropped to less than 500 000 individuals by 1970s (Hamr 2016). The primary cause for this decline was habitat destruction and market and subsistence hunting (Fricke 2008). Active market hunting for meat and hides was also a prime cause of American Bison population decline (Gates 2010). An estimated number of 60 million American buffalo (Lott 2003) dropped to 168,000 by 1800 (Gates 2010). It is crucial for market hunting not to be confused with trophy hunting, as system of trophy hunting is what facilitated the recovery of these species in later years.

Back in the 1800s, in contrast to farmers, the urban residents had more of one thing than those working the field – time. Time that they very much intended to fill with activities like hunting, which even in spite of its rigours and challenges, they found to be appealing. The conflict soon arose between the market hunters, who profited from wildlife, and sport hunters who highly valued wildlife and pushed for its responsible utilization. These were the sport hunters who would establish first wildlife refuges (Carroll's Island Club 1832, Gunpowder River in Maryland; Trefethen 1975) and develop laws to protect game and non-game species from exploitation (Organ et al 2012).

A well-known and a highly influential representative of these sport hunters was George Bird Grinnell, a Yale-educated naturalist who in 1879 acquired the sporting journal *Forest and Stream* and turned it into call for wildlife conservation, which many would follow. In 1885 Grinnell reviewed a book authored by Theodore Roosevelt about his hunting exploits in the Dakotas. He praised the book in his review, however also outlined some of the inaccuracies presented in the work of Roosevelt. Upon meeting, the two agreed that big game had changed and experienced dramatic declines in the past decade, their many conversations and discussions led them, in 1887, to establish the Boone and Crockett Club (Organ et al 2012) – the club that took the responsibility upon itself to recover these endangered species.

The organization recognized that to facilitate species recovery, and to prevent future threats to populations, one of the most important things to encourage was a sustainable harvest. This system protects the core of wild breeding population, specifically, females and young male individuals, while focusing the harvest on older males. These are the individuals who have already reproduced, and have already passed on their genes (conservation visions. 2017). As a result of trophy hunting combined with other

26

conservations tools, North America's critically endangered wildlife species recovered to more stable numbers.

Some organized groups are demanding a ban on trophy hunting as the activity is regarded unethical and claimed to be ineffective. However, the consequences of eliminating one of the most crucial conservation tools are largely underestimated. In response to poaching and illegal Ivory trade, Kenya passed a hunting ban in 1977. It was believed that with hunting pressure off, the game would return to high numbers. What took place as a result of the ban was quite the opposite of its intent. Charismatic megafauna – lions, rhinos, elephants and large antelopes – are experiencing extreme declines since late 1970s. Wildlife numbers have declined on average by 68% between 1977 and 2016. Giraffe with an estimated population size of 76,236 in 1977 dropped to 25 193 by 2013. Number of buffalo decreased to 40,245 by 2013 from 66,169 in 1977. Impala and hartebeest experienced 84% decline in population size while the eland and oryx numbers were reduced by 78% since the year hunting ban was passed. Warthog, lesser kudu, Thomson's gazelle, topi, elephant, Grevy's zebra and waterbuck are other unique species undergoing extreme population declines. These declines raise serious concerns about the future of African wildlife, in particular the effectiveness of wildlife conservation policies, strategies and practices in Kenya.

Of course, it is false to assume that banning consumptive use of wildlife is the sole reason for such declines, other causes of negative wildlife population trends include exponential human population growth, habitat loss, increasing livestock numbers, declining rainfall and a striking rise in temperatures, yet, the fundamental cause remains to be ineffective policy, institutional and market failures (Ogutu 2016).

7. Predator Management.

Habitat quality, food sources, sex ratio, climate, diseases – these are the factors influencing population size of any wild species, however predator numbers also play a crucial role in determining the abundance of prey.

One of the reasons species like deer, hare, pheasants and such others still exist where they do, is to some extent due to predator population management. The relationship between predators and prey was always known to conservationists and has helped in regulating many problematic species. However, the predator-prey relationship is quite complex and dependent on more than one variable.

There are two approaches used when it comes to predator role in ecosystem management – predator-driven (top down approach) or prey-driven (bottom up approach). The idea of top down approach is that when predators are eliminated from the area, prey species are granted the opportunity to survive and reproduce, while the idea behind bottom up approach lies in modification of habitat in favor of preys, for example, more cover means more chance the prey has to successfully hide form the predator.

However, the issue is far more complex and goes deeper in theory. In some ecosystems, certain predators are classified as *Keystone species,* which recognizes that these predators, once removed, will have a tremendous effect on more than one species either by direct or indirect means. If coyotes for example, are removed from a site for long enough period of time the deer fawn survival will increase, however, the elimination of coyotes may also lead to population increase of nest predators such as grey foxes and feral cats, which cause population declines in quail. Therefore, the removal of coyotes will have a direct influence of deer fawns while having an indirect influence on quails. It is important in this case to manage coyote population in such ways that neither deer nor quail is put at risk.

It comes with no surprise that today human activities have significantly disturbed this natural cycle. Unsustainable killing and trapping of predators, poaching, and introduction of non-native species around the world led to the establishment of invasive predator populations which are not only causing damage to the native ecosystems but also to the human settlements nearby.

Because of their generalist behaviors, invasive predators most frequently outcompete the native species as they are highly adaptable and can easily withstand different climatic conditions. Invasive predator species either occupy an available niche or outcompete a previous inhabitant of the niche. Domestic cats proved to be successful in this regard as they managed to reproduce, disperse and outcompete other predator species in no time (Marra & Santella, 2016). Based on the experts' estimates 14 billion birds, mammals, reptiles and amphibians are killed in the United States annually by free-roaming cats (Adler 2014). Managing predators in order to safeguard more vulnerable prey species becomes of utmost importance.

Predator population management is no different than any other species management. The targeted populations are first monitored by wildlife managers, after which the most suitable management goals are determined for each population. Depending on the interests of the public as well as conservation, the predator populations are increased, decreased or kept stable. Once the objective is clear wildlife managers set harvest regulations (quotas, hunting seasons etc.). Under most circumstances, even if the current objective is to cause decrease in predator population, the overall goal is to prevent population declines of every species over time.

In cases where furbearer populations are causing significant damage, either by causing direct threat to endangered species or damaging fish and wildlife habitats, it may be desirable to reduce their populations. The need to lower carnivore populations is also caused when they are creating hardship for landowners or agricultural producers. In such circumstances, hunting and trapping regulations are adjusted, so that harvest exceeds surplus production. The harvest is then added to the annual mortality and it is this *controlled additive mortality* that ultimately causes the population decline (The Northeast Furbearer Resources Technical Committee, 2015).

On the contrary, when the objective is to recover endangered species or after population was severely weakened, it is naturally, desirable to increase the population size. In situations like these harvest is restricted or prohibited for as long as it is need for the populations to grow to more stable numbers.

Beavers are good example of a furbearer that requires intensive management. On one hand, wetlands created by beavers are extremely productive which harbour a wide variety of plants and invertebrates, thus providing habitat for hundreds of fish and wildlife species. It is important to keep beaver populations stable not to jeopardize species that depend on beaver-made habitats. However, beaver populations require constant control in order to reduce conflict with humans. The problems with beavers flooding roads and damaging private property are common issues.

Previously, predators like wolves in Europe for example, were viewed as pests because of the tremendous damage they brought to game and livestock and thus, the attempt to eradicate them completely was done on several occasions. In Poland's Białowieża Forest the wolves kill around 72 deer, 31 wild boars and 15 roe deer per 100km^2 every year. However, according to the diet analysis the impact of wolves in commercial forests are much bigger compared to Białowieża Forest (Jędrzejewski et al. 2002). Wolves are highly effective predators who are capable of reducing the density or deer species and slowing down the population growth rate in herbivores, preventing them from reaching maximum densities set by food resources (Jędrzejewska and Jędrzejewski 2001).

Woodland caribou is also on a decline across Canada and it is due to human-induced changes to food webs which results in increased predation thus higher caribou mortality.

After 12 year study in Alberta Canada, wolf removal translated to a 4.6% increase in mean population growth rate of the Little Smoky population mostly through improvements in calf recruitment (Hervieux. 2014).

In Białowieża Forest of Poland, lynxes were responsible for killing from 110 to 181 roe deer per 100 km^2 annually, comprising 21-36% (average 26%) of the spring population. Such degree of predation caused a significant decrease in roe deer population growth which expressed in numbers is 64% on average. Moreover, lynx does not show preferences with respect to sex, age or physical state of a prey. Thus, lynx was the factor most influencing the population size of roe deer. (Okarma et al. 1997).

The degree to which the lynx affects red deer is smaller compared to that of roe deer, however, it still has noticeable influence on the red deer populations. According to the same study, lynx was killing 42 to 70 deer per 100 km^2, comprising 10% of spring population on average. As a consequence deer population growth was reduced by 21-43% (average 33%). When it comes to red deer, lynx shows a clear preference for juveniles and individuals in physically weak conditions (Okarma et al. 1997). While devastating to deer, lynx was playing an important role in proving food for number of other species. Remains of the prey are normally used by other mammal and birds who show generalist feeding behavior (Selva et al. 2005).

The whole point of presenting these numbers is to show the high effect predators have on ungulate species. To ensure the safety to all species, it remains of utmost importance not only to manage their habitats, but to control predator populations. Yet, when it comes to predator control the methods remain to be somewhat debatable. Fencing, habitat modification, trapping and shooting are few amongst many methods that are used by wildlife managers to manage carnivore populations. It is important to point out that the effectiveness and cost associated with a specific method will determine its appropriateness for a given situation – it is incorrect to assume that one method will work in every situation.

In attempts to be as ethical and humane as possible, methods like relocation, and chemical birth control are adopted instead of hunting and trapping, yet they usually produce little effect. Past studies have concluded that hormone injections or implants could serve as successful means of reproduction control. However, challenge lies in that each animals is required to be treated individually, and in case of hormone injections the procedure has to be repeated multiple times. Consequently, such treatment is not only expensive but is also incredibly difficult to apply to large animal populations.

Certain animal welfare and wildlife agencies are supporting the idea of *immunocontraception* as a possible solution to regulate reproduction and thus population size of certain predators. Immunocontraceptions are said to be inexpensive, easy to administer and long lasting as they use vaccines that target specific hormones or reproductive tissues. However, this research is still at its beginner stage and the final product will take years more to be available on the market. Regardless, some technical problems still remain when it comes to such method. Firstly, animals have to be vaccinated individually in order to ensure safe and effective administration. Secondly, dart guns and blow guns which the vaccine is delivered with have very short shooting range so it is necessary to get as close to the animal as possible which is a difficult mission in itself (Muller et al 1997). Thirdly, specific percentage of population has to be sterilized in order to cause population decline over time. In case of feral cats, 70-90% of population must be trapped and spayed/neutered to cause even a slight population decline (Foley et. Al 2005), such high trapping success is not only improbable but next to impossible. Lastly, even if the population was successfully sterilized and their decline is ensured, that still does not solve immediate problems. Spayed and neutered feral cats will no longer reproduce but will still prey on small birds and mammals further endangering their populations. Sterile beavers will no longer reproduce, but will still process functioning teeth and will therefore continue cutting trees and building dams.

The more successful and by far the more effective method of predator control proved to be hunting and trapping. Shooting as the means of control can be a highly selective method, given of course that the hunter is able to correctly identify a species and is positioned for an accurate killing shot. Trapping is another method used for predator control and like shooting, trapping effectiveness depends strongly on the experience and skills of a trapper. It is important not only to correctly identify the species but to choose the most suitable trap in order to avoid any damage to the animal. Placing a trap is also something that every trapper knows how to do well. Traps are placed on the route that is most often used by individual animals. Choosing the correct location is also important as to avoid catching non-target species. Trapping has proved to be successful on many occasions (The Northeast Furbearer Resources Technical Committee, 2015).

Islands along the Maine coast provide critical habitats for colonial-nesting seabirds. Atlantic puffin, Razorbill, Great Cormorant, and Arctic Tern are some among many other species that rely on these habitats to maintain their populations. In fact, the Maine Legislature recognized many of the sea bird nesting islands as "Significant Wildlife Habitat", an indication of the high conservation value of these nesting islands.

Recently however, mammalian predators like river otter and mink have made their way to these islands during the time when adult birds were rearing the chicks. Such time of arrival resulted in exceptionally high chick and adult mortality. Unlike when predators arrive during an incubation stage, when adults abandon the eggs and the island for the whole season, during rearing stage parents are committed to raising their chicks and will not leave the island. At National Audubon Society-owned Stratton Island, otter and mink ended up killing more than 500 terns in less than a week (The Northeast Furbearer Resources Technical Committee, 2015).

In 1970s in an effort to restore and protect nesting sea bird populations, U.S. Fish and Wildlife Service's Maine Costal Islands National Wildlife Refuge (NWR) was established. NWR recognized that the factors limiting seabird population growth was not only habitat degradation, human disturbance and food resource availability but also avian and mammalian predator management. Between 2007 and 2014, NWR personnel managed to trap 14 mink on Western and Eastern Brother Islands where Common terns, Common Eider, Leach's Storm-petrels and Black Guillemot are found nesting. These birds are very vulnerable to mink predation and fall as easy prey. Black Guillemot is a species that lays only one egg a year and will not renest in that year if its nest was destroyed. Because of active trapping wildlife managers successfully maintained an average of 350 nesting pairs of seabirds over the course of the management period (The Northeast Furbearer Resources Technical Committee, 2015).

Hunting and trapping have played, and are playing significant role in predator management. These activities that the public rejects so strongly have contributed to the wellbeing of the wildlife ecosystems safeguarding species that otherwise would have gone extinct.

8. Case Study: Methods of Feral and Stray Cat Population Control: Trap Neuter Return vs. Trap and Remove

Feral and stray cats pose grave danger to wildlife across the globe. It is estimated that the number of free-ranging cats exceeds 100 million in the United States which has by far the largest cat population compared to any other nation (Marra & Santella, 2016). This is a matter of great concern as around 14 billion birds, mammals, reptiles and amphibians are killed in the US annually by free-roaming cats. These small felines are responsible for population decline, reduction in geographic distribution or extinction, in 175 species of reptiles, birds and mammals on 120 different islands around the globe (Medina, et al., 2011). Cat colonies have proved to have a major effect on composition of some of the rarest species which raises big concern to conservationists. These are the animals – our cat companions – that were directly involved in extinction of 33 bird species among which are Stephens Island wren *(Xenicus lyalli)* of New Zealand and Socorro dove *(Zenaida graysoni)* of Mexico. Coming up with a method to control the overpopulated cats, today is of extreme importance.

When considering the different control means used to regulate cat populations there are two methods that are most talked about. *Trap Neuter Release* (TNR) and *Trap and Remove* are methods exercised today, both of which are highly controversial issues.

On one hand TNR seems like a more humane method that preserves lives of cats. TNR advocates claim, that the treatment frees cats from the stress of mating, pregnancy and parenting. Behaviours associated with unaltered cats, such as yowling and marking the territory with urine, will disappear immediately after the treatment[6]. Without having kittens to care for, adult cats will have more available resources and thus better physical health. It is also believed that physical health of cats is further improved by vaccinations given against distemper, herpes, rabies, feline leukemia and other diseases (Marra & Santella, 2016). Additionally, male cats will slowly loose the urge to roam and fight, reducing the risk of injury.

However, there are science based arguments that point out the certain limitations of TNR. It is important to follow and monitor the treated cats in order to understand the behavioural changes. While they may be relieved from mating, pregnancy and parenting stress they are not in any way relieved of their natural instincts to kill. Returned cats are sure to go back to killing and hunting small mammals and birds endangering their population status. From the point of view of ecology and conservation this is unacceptable.

---

[6] http://www.houstontx.gov/barc/trap_neuter_return.html

While the Humane Society of the United States believes that outdoor/unowned cats deserve and are entitled to protection, it also realizes that other wildlife species are under threat due to these cats. As a result, HSUS has grown to become strong TNR advocate[7]. BARC animal shelter is a strong advocate for TNR as well, and argues that although "Trap and Remove" might work on short term it has no effect over time. Hunting and trapping is believed to be very selective method but also a challenging one. It is difficult to target an animal that is small, shy and nocturnal (Orueta & Ramos, 2001). BARC argues that colonies subject to "Trap and Remove" program will ultimately end up increasing their number to the original size due to what is known as the vacuum effect[8]. Alley Cat Allies shares this view and calls the practice of Catch and Kill pointless. Removing feral cat populations will open up the habitat, and draw new cats to the area either from neighbouring territories or born from survivors. Due to the vacuum effect the population will return to the original number sooner or later, drawing the community into a costly and endless cycle of trapping and killing (Alley cat Allies & , 2011).

The vacuum effect is based on the idea that territorial animals residing in high-quality sites will outcompete certain individuals in that area which are weaker and inferior. This situation will result in "winners and losers" in which the excluded animals are forced to occupy nearby territories without appropriate resources. This circumstance creates "open spots" in the original population and sooner or later, attract other members of the same species from the neighbouring areas who will move in to take advantage of the same resources that attracted the first group. TNR advocates claim that because of the vacuum effect there is a never-ending supply of cats to be attracted to the preferred areas when cats are permanently removed (Alley cat Allies & , 2011). Given this assumption, removing cats is a pointless practice if the main objective is to reduce the overall population of the cats. However, there are few problems regarding this claim. First of all,

---

[7]www.humanesociety.org/animals/cats/facts/cat_statement.html?credit=web_id83574224#cats_and_wildlife.

[8] http://www.houstontx.gov/barc/trap_neuter_return.html

36

the notion of infinite cat supply is non-existent, rather if cats continue to appear it is probably due to feeding stations set up by humans. Secondly, the vacuum effect that presumably draws new cats into the colony when members are trapped and removed, will also be at work when a member of the TNR colony dies. And finally, and most importantly, domestic cats are not territorial species. (Marra & Santella, 2016).

A study done in Israel in 2011 examined four feeding groups of stray cats, two of which were spayed/neutered, and two of which remained sexually intact. Over the course of one year study, the number of adult cats in two neutered groups increased considerably because of higher immigration and lower emigration rates. Compared to the neutered groups, the individual cat numbers in the unneutered groups showed a decline. Kitten survival in the treated groups was significantly higher than in the unneutered groups as well. The study concluded, that there are two major changes in the dynamics due to the TNR treatment. Firstly, sexually intact cats immigrate into the neutered groups more readily, and secondly, neutered cats reduce their emigration rates, possibly because of a reduction in reproductive and competitive pressures (Gunther, Finkler, & Terkel, 2011). In this case, the vacuum effect increased the number of cats in existing TNR colonies. This will not always be the result. Vacuum effect will take place at times, but not always. The degree of replacement depends upon number of factors, such as number of cats in the surrounding area, the rate at which those cats are reproducing, the number of cats that are behaviourally dominant, as well as on the abandonment rate (Marra & Santella, 2016). In order to maintain a high proportion of neutered individuals in such cat colonies, highly intense TNR campaigns are therefore necessary.

In a 2014 report issued by Feral Cat Task Force, method of TNR is taken a step further. FCTF argues that it is crucial to manage the colonies after releasing cats back to their original habitat. Under the management program, stray and feral cats will be impounded and returned to their owners in case they have a registration or identification number. FCTF emphasizes that TNR cat colonies must be rigorously registered, certified and monitored. Colonies must be properly maintained with a minimum 90% spay-neuter rate. Sick and/or injured individuals must be removed; new arrivals and new litters of kittens would be taken and made available for adoption or euthanasia (Adler, 2014).

In theory and when read on paper TNR method offers the ideal solution to the problem. It conserves the lives of both cats and birds thus solving the problem of biodiversity loss. However, if investigated further it can be seen that the problem is not solved but made worse.

TNR Fails due to many reasons but probably the most important and obvious one is failure to trap (Marra & Santella, 2016). The fraction of cats that must be spayed and neutered to force population decline is somewhere between 71% and 90% (Foley, Foley, Levy , & Paik, 2005). To achieve such trapping success on the field is not only impossible but highly improbable. Since adapting the method of TNR this level of trapping and neutering has not been documented (Marra & Santella, 2016). On top of this it is safe to assume that cat colonies are constantly receiving new individuals which further increases the population size and increases the difficulty of trapping and neutering.

Results of a study done by Andersen and Martin (2004) showed that a 50% humane euthanasia rate would yield a reduction in the feral cat population by 10% per year; these results are quite impressive compared to method of TNR where even 75% spay/neuter rate will lead to increasing feral cat population (Andersen, Martin, & Roemer, 2004). According to the same model at least 88% of feral cat population must go through the TNR treatment to merely stabilize population growth.

Furthermore, TNR program claims to vaccinate feral and stray cats against the diseases, however, in reality very few cats receive boosters and it makes cats vulnerable to diseases. Once subjected to a disease it can be easily passed on to other cats, wild animals and humans (Marra & Santella, 2016). On average spaying a female cat takes 4-

6 minutes while neutering a male last for less than 3 minutes. TNR program is offered by non-profit humane societies to city, or by country run animal shelters to private veterinary practices. While TNR procedure is relatively fast and accessible it is quite expensive and requires a trained veterinarian to perform the surgery properly. Prices at Oregon Humane Society for spaying a female range from $40.00 to $42.50 while for neutering from $30.00 to $32.50 (Marra & Santella, 2016).

One of the key arguments for success of TNR is that over time feral and stray cat population will decrease and eventually stabilize to a number that is no longer harmful to biodiversity. However, it can be argued otherwise. Even though the number of original colony members may decrease, illegal abandonment of unwanted cats will prevent the colonies from decreasing. It is a fact that human interference plays a significant role in population dynamics and feral cat colonies are no exception. Abandonment of unwanted cats remains to be a serious problem around the world (Castillo & Clarke, 2003). In Spain in 2008, 38 631 cats were admitted to shelters while 35,794 in 2009, 35,983 in 2010, 33,851 in 2012, and 33,532 in 2013 (Fatjó, et al., 2015). In 2010 in UK 156 826 cats were cared for in shelters (Stavisky , 2012). These data gives a rough rate of cats abandonment and shows how much of a problem abandonment is.

There have been numerous attempts to control cat populations using different methods, some of them which proved successful. In 1949 a pregnant female was introduced to 30,000ha Marion island. By 1975 cats were killing 450 000 burrowing petrels per year. In 1977 the cat population reached 3 400, causing local extinction of a petrel species by 1980. In order to deal with this problem Feline panleucopanenia virus (FPV) was introduced in 1977 in South Africa. This caused reduction in cat numbers to 615 by 1982. Combined with FPV night hunting decreased the population to 89 by 1989. The remaining cats were eliminated by trapping and hunting. Since 1992 no cats have been seen in the area which recovered almost extinct bird populations to more stable numbers (Huntley, 2001).

The most common and the most successful eradication program on 91% from a total of 43 islands was trapping and hunting. Most of the times these programs would be used in combination. Other successful methods included poisoning which is legal and practiced in many countries today. Globally, feral cats have been removed from at least 48 islands: 16 in Baja California (Mexico), 10 in New Zealand, 5 in Australia, 4 in the Pacific Ocean, 4 in Seychelles, 3 in the sub-Antarctic, 3 in Macaronesia (Atlantic Ocean), 2 in Mauritius, and 1 in the Caribbean. The majority of these islands are small. The largest successful eradication campaign took place on Marion Island which was 290 km^2 in size. (Nogales, et al., 2004).

Management approach of stray cats differs among nations. Albania, Armenia, Azerbaijan Republic, Bulgaria, Moldova and the Ukraine reported to be culling stray cats. Belgium and Greece were the only two countries that actively practiced Trap Neuter Return (TNR) programs as the means to control feral and stray populations. However, majority of the European countries reported to be using combination of different methods; mostly TNR and catch-remove. Belarus, Estonia, Germany, Lithuania, Norway and Portugal reported to be catching stay cat but not practicing TNR or culling methods (Tasker, 2007).

Talking about permanent cat control is a challenge, as it is natural for social and ethical issues to arise from the general public. In Some countries like Great Britain feral and stray cats are illegal to be shot or poisoned since they are regarded as domestic animals (Neville, 1989). It is worth emphasizing that feral cats' behaviour is nothing like of domestic cats, feral cats show more resemblance to the wildcat *(Felis sylvestris)* as they never had any contact with humans and thus lack trust towards them. In the United Kingdom cats have to be trapped using a humane method and put down by a lethal anaesthesia by professional veterinarians (Neville, 1989) .

This raises interesting ethical issues. Is it ethical and just to ensure the survival of one species knowing that other species are at risk of extinction? Who should we prioritize?

Should we allow the predator to do what it naturally does? Or should we side with more vulnerable species?

Sometimes in conservation, professionals are faced with difficult decisions. At times the method that may seem the most "humane" like TNR has the outcome opposite of its intent. Hunting and trapping do not seem very humane to many people, however, sustainable hunting leaves us with the result of a healthy and thriving populations.

9. Human-Wildlife Conflict

Conflict arising between rural communities and wildlife is the most widespread, and one of the most intractable issues in conservation biology. Agricultural lands are frequently damaged by wild animal species causing farmers great economic losses. This issue does not only concern the agricultural damage, as human-wildlife conflict ranges from grain-eating rodents to man-eating tigers. Coexisting with such species can impose a variety of significant costs upon local people, including depredation of livestock, crop damage, destruction of stored food, attacks on humans, and/or disease transmission from wildlife to livestock or humans.

In US agricultural producers' survey, 89 percent of the respondents admitted to be experiencing problems with wildlife species. Wild animals seem to damage private property of 80 percent of the respondents, out of which 54% claim that the losses exceed $500 every year. Wildlife sanctuaries are crucial to conserve populations and protect them from predation, poaching and other dangers, however, due to the overwhelming damage brought by wildlife, 40% of all agricultural producers begin to oppose the creation of such establishments. At the same time, 26 percent of the respondents admit, that this extent of wildlife damage reduces their will to support wildlife habitats on their property. At this stage, wildlife starts to lose its value and becomes a pest that farmers try to get rid of. This kind of attitude in the rural community is worrying. Agricultural lands are farmers' main source of income, therefore, in attempts to protect the land from wildlife damage, locals are forced to turn to management means, that are unsustainable and ineffective on long-term. These methods include illegal killing and trapping, that puts wildlife populations in danger (Messmer 2000).

The fact, that these damages still occurred after annual expenditure of over 40 hours and $1000 per farmer to prevent the damage, is an indicator that a more active control method of wildlife species is required. It was estimated that losses on agricultural lands exceeded $2 billion despite the fact that over 91 million hours and $2 billion was spent to prevent the damage. Of these losses $160 million was to livestock and poultry; $53 million was to vegetable, fruit and nuts; while $30 million worth of stored crops were destroyed.

Agricultural industry is not the only industry that suffers damage, beaver *(Castor Canadensis)* and deer *(Odocoileus spp)* are costing timber industry devastating damage as well. Estimated loss to tree plantations due to beaver activity was estimated to exceed $22 million in 1995. Damage caused by deer cost $367 million to the timber industry (Messmer 2000).

White-tailed Deer *(Odocoileus virginianus)* have especially strong effect on altering plant species composition and distribution. For example, deer have an ability to change density of legume species, which shelters nitrogen producing bacteria. The change in their populations greatly reduces nitrogen content of the soil, thus, influencing agricultural production success (Russell, 2001). Deer feeding behaviour alters the success of certain plant species, consequently, they may also change the success or distribution of other herbivore species.

Woodpigeons *(Columba palumbus)* are major agricultural pests that bring tremendous damage to oilseed rape crops. It was estimated that without special control programs, damage would amount to £45 million in East Anglia alone. In order to avoid such large scale damage, more that 50% of shooters are involved in protecting crops from pigeons and other avian pests (PACEC 2014).

It is important to point out that apart from damages on private property (in both rural and urban areas) wildlife populations also pose danger to the residents in areas where their population numbers are unregulated. Human-wildlife conflicts include transmittable illnesses, attacks, deer-vehicle collisions and bird-aircraft strikes, all of which might be fatal. More than 5000 people are injured or hospitalized while more than 400 people die due to wildlife related incidents (Messmer 2000). Most of these economic and ecological consequences can be either avoided or decreased if wild animal numbers are reduced to levels that are more compatible with human activity and land use.

It is important to understand, that wildlife conservation does not only encompass animals. Conservations is an extremely complex field of study, which considers the wellbeing of humans to be as important as the wellbeing of animals. Favouring one factor over the other is not sustainable and will not lead to positive outcomes over a long-term periods. Hunting and trapping combined with other management methods offers the only sustainable solution to the issue of human-wildlife conflict. If wildlife is left unmanaged not only will it destroy its own habitat but will also lose its value in the eyes of humans.

10. Driving the economy

With more than 7 million hunters in Europe (FACE. 2010), and 11 million in USA (US fish and Wildlife Service. 2017), and with more than €16 billion generated by trophy hunting in Europe alone (Middleton 2014), hunting becomes the driving force for wildlife conservation and sustainable development. Hunters around the world are making contributions to all of the main sectors of the economy by both direct and indirect means. For example, farmers are compensated for crop damage in the primary sector, hunting gear is purchased from the second sector, while tourism services are paid by hunters in the third sector (FACE 2014).

Hunts and trapping licenses are usually purchased by people who are willing to pay substantial amounts of money for the experience. As a result, revenue generated by hunters provide management organizations with funding for research and conservations efforts to protect wildlife.

Across Europe, revenue generated by hunters directly benefits both wildlife and rural communities. In 2007 in Ireland, hunting ended up generating €111.6 million (FACE. 2014), 82 percent of which was spent on developing the rural areas, 16 percent went to developing cities and larger towns, while the remaining 2 percent was spent outside the country (Scallan. 2012). In the UK, at least 600,000 people shoot live quarry, clay pigeons or targets, these activities are contributing £2 billion to the UK's economy. As a result, two million hectares are being actively managed for conservation. Shooters spend almost 4 million work days on conservation programs, this is an equivalent of 16,000 full-time jobs. Moreover, shooting providers spend nearly £250 million a year on conservation while more than £2.5 billion is spend by hunters on goods and services each year. It is crucial to point out the fact, that shooting and hunting support 74,000 full-time jobs, thus strongly reducing unemployment (PACEC. 2014).

In Italy €3.26 Billion is estimated to be generated by 850,000 official hunters, while at the same time creating in total of 43,000 jobs. In 2008 in Finland, around 40,000 hunters took part in voluntary labour-intensive activities, ranging from game monitoring to assisting with assisting works related to road accidents involving wildlife, these programs were estimated to value €7.1 million (PACEC. 2014). In Greece hunters' yearly contributions finances 400 game guard who are involved in tracking and tackling illegal activities, which annually amounts to €7 million (Papadodimas. 2011).

In 2016, 11.5 million people participated in hunting activity in the United States. Hunters spend up to 20 days on the field pursuing wild game. US's most popular big game like deer, elk, and wild turkey attracted more than 9 million hunters (80%) while small game including squirrels, rabbits, quails, and pheasants ended up attracting 3.4 million hunters (31%). Geese, ducks and doves attracted 2.4 million hunters (21%) who, when combined in total, spend more than 15 million days on the field. Other animals like coyotes, groundhogs and raccoons were popular game species for 1.3 million hunters (11%), spending 13 million days hunting. Overall, in 2016 hunters in the United States ended up spending $25.6 billion on trips, equipment and licenses with an average expenditure of $2.237 per hunter (US fish and Wildlife Service. 2017).

Between 2000 and 2008, in seven SADC (South African Development Community) countries (Botswana, South Africa, Namibia, Zambia, Mozambique, Zimbabwe and Tanzania), trophy hunting ended up generating over US$190 million per year. Sport hunting in many African countries, like South Africa, is contributing more than 68.3 million USD to the gross income. It is this revenue that drives sustainable development in the regions by funding and building facilities crucial for national advancement. Tourism sector in Tanzania plays a key role in foreign exchange earnings, and contributes more than 50% to total export earnings. Tourism of Tanzania is estimated to directly support at least 30,000 jobs on the mainland and 6,000 more in Zanzibar with wildlife safaris being primary attraction especially in the Northern Circuits (Vernon. 2010).

11. Game Meat vs Produced Meat

History of hunting

It is likely that meat entered our ancestors' diet as scavenged food, either through stealing from other predators, or collected in the aftermath of natural disasters like fires and flooding. Our ancestors would have started hunting in a very primitive way, using tools that would hardly classify as weapons. At that point, incapable and inexperienced early humans took advantage of weak, sick and old individuals. Turning animal protein as a routine food rather than an occasional dietary treat allowed the rapid growth of the human brain. Proteins and nutrient that come from cooking and eating the meat played a deciding role in the evolution of the early humans, eventually turning us into highly intelligent animals that we are today (Milton 2003).

To support and feed this rapidly developing brain, humans needed more of these proteins and nutrients, which in turn necessitated more efficient and better hunting efforts – this was made possible by our increasing brain capacity and the intelligence that came with it. Gradually the methods and tactics of hunting became more sophisticated, we started to carve better tools and form more complex social relationships which increased the probability of a successful kill. This culture of group hunting developed skills and qualities such as cooperation, better communication, and ones again made humans realize the value and the importance of *sharing* among the group. Even today, these are the qualities highly valued in modern day society and are qualities that are strongly associated with hunting community.

Introduction of meat into our diet played a crucial role in the human evolution and shifted our relationship with the wild animals. It comes as a surprise when an intelligence, coming from an evolved brain, powered by minerals and protein from wild meat will eventually speak against hunting and consumption of meat all together.

Food security

Unlike our ancestors for whom, in the early stages, meat was an occasional treat, we have managed to develop farms that produce meat in exceptionally high numbers in short periods of time. Despite the fact that livestock sector is able to feed millions of people, such easy and comfortable lifestyle comes at a certain price. The livestock sector is known for its high contribution to global climate change. It is responsible for 18% of anthropogenic greenhouse gas emissions as well as 37% of anthropogenic methane emissions. This amount of methane has more than 20 times the global warming potential of CO_2. The emission of nitrous oxide adds up to 65%, while the emission of ammonia is around 68% worldwide (Bhat. 2014).

The question at hand is – is it possible to obtain meat, and feed millions of people without polluting the environment as a result of massive land use?

Compared to other nations, United States is among those with highest food security. 114.9 million households were food secure throughout 2018 – this is about 88.9% of the whole population. While this figure is worth celebrating, it is important to be aware that the remaining 11.1% (14.3 million) of American households face food insecurity, 4 to 7% of these households had *low* to *very low* food security some time during 2018[9].

Hunting has the ability to address the issue of food insecurity. An important aspect to note is that the harvested meat is never consumed solely by the hunter. This food is shared with family members and friends. Some of the meat is even donated to programs that aim to provide high quality foods to ones in need. *Tennessee's Hunters for the Hungry* program has donated about 4 million meals to Tennesseans in need. About 147,000 pounds of wild meat was donated in the 2016-2017 season, enough for 588,000 quarter-pound servings. In 2015 *Virginia's Hunters for the Hungry* provided 301,809 pounds of venison to charity, which is enough for 1.2 million quarter-pound servings. Since the launch of the program in 1991, 6.38 million pounds of venison have been donated, providing more than 25.5 million servings. *Texas Hunters for the Hungry* has donated about 9 million meals since it started, while 1.8 million meals were donated from *Alabama's Hunters Feeding the Hungry*. The notion of sharing is important and well recognized among hunters[10].

---

[9]https://www.ers.usda.gov/topics/food-nutrition-assistance/food-security-in-the-us/key-statistics-graphics.aspx

[10] https://www.academy.com/explore/how-hunters-are-feeding-hungry-america

Nutritional value

It is no surprise that game meat is becoming more popular food choice because of its high nutritional value and health benefits. Because of its quality, non-hunters are starting to appreciate hunting more and more as an activity and come to value hunters' efforts. A 2009 study randomly surveyed 1,067 Swedish residents, in order to test the association between non-hunters' frequency of game-meat consumption, and their attitudes toward hunting. The study concluded that game meat was consumed at least once a year in 65% of non-hunters' household. Moreover, the same study concluded that 80% of non-hunters expressed favourable attitudes toward hunting (Ljung et al. 2012).

Hunting is also valued in East Europe. The demand on wild boar meat is on the rise in Hungary as well. The sales increased by 86% since 2006, reaching its peak in 2012 when 5.7 million kg of meat was sold to the public. Sale in red deer meat is also experiencing a rapid rise, which saw 71.3% increase between 2006 and 2016. The rises in sales of roe deer, fallow deer and mouflon meat are no smaller than that of red deer and wild boar. From 2006 until 2016 sales have risen by 53.6%, 86.6% and 47.1% (Komarek 2019).

Wild harvested meat serves as a better alternative to domestic animal meat. Because of the natural lifestyle, eating habits, and quality of their food and habitat, meat of the game is significantly different in its nutritive value compared to the meat of farm animals. Game meat is produced without the use of any antibiotics and hormone injections, therefore, it can be considered as an "organic product", which is what most people seek in the markets.

As opposed to domesticated animals whose diet consists of mostly grain and corn, and are kept in closed crowded circumstances with minimal space to move, wild game live freely in their natural habitats, leading more active life and eating a natural diet. Consequently, wild meat tends to have lower fat content compared to non-game meat. Depending on the age, sex, and the nutritional status of the animal, fat content ranges from 1 to 6% only (Komarek 2019).

Out of all types of meat, it is game-meat that has the highest protein content, which amounts to 21-25% (Komarek. 2019). Another one of its advantages is its high mineral and vitamin content. Wild meat is an important source of B vitamins (B1, B2, B6 and B12) as well as vitamins A and D. Game meat is dark red in color which is the sign of high blood and myoglobin content. It is rich in niacin, iron, zinc, and phosphorus (Nilson. 2018).

Ethics

We have long reached the stage of development where we are no longer concerned about our day to day survival. We now have the luxury to give our mind to more intellectual matters like philosophy, morality and ethics. We think about what is fair and what is just. We recognize the rights of the minorities and come to respect them. We claimed the responsibility of representing those who cannot speak. We now empathize with animals and in some cases come to see them as equals.

It is impossible to talk about hunting and game meat without having to mention, and the increasing human concern around meat consumption. Animal rights movements are gaining influence and power very rapidly. Choosing veganism and vegetarianism as a lifestyle has become an attractive idea, and has gained popularity in our modern society. It is often that animal rights and welfare activists point out the practices that take place in meat industries. They emphasize the violent methods used against individual animals, they speak against limited freedom, suffering and mistreatment of the living beings that are raised solely for their meat and products. The amount of antibiotics and other drugs administered to domestic animals, and large scale use of pesticides and herbicides for growing feed crops, greatly alarm people who are concerned with their personal health. Meat industry also worries those who care for the environment and sustainable use of the environment. Their arguments against meat eating circle around wildlife habitat degradation, biodiversity loss, and degeneration of ecosystem services. 30% of the total land surface is used for livestock production, 33% of arable land is used to grow feed crops for livestock while 26% is being used for grazing (Bhat. 2014). The amount of land and water that is required to raise food, not for human consumption, but for livestock raises great concern in people.

With the emergence of agriculture, aquaculture, and meat industry, people are no longer involved in personally investing time and energy in acquisition of their own food. This highly urbanized world, dominated with modern technology results in upbringing of the society that has become detached from not only producing their food, but from natural world itself. Our meat is now raised, treated, killed, processed and preserved for us, in some cases it is even cooked for us.

The question is raised – is it possible to obtain meat without the massive land use, without any chemical treatment, without torture and violence? The answer is yes. And 18 million people in United States and Europe have been doing just that.

Hunting, and hunters are very often demonized, shamed, called names, and accused of evil doing, yet, the wild meat that hunters harvest is produced without any type of land development and is taken with no damage brought to the habitat. The majority of people who are engaged in hunts are committed to *fair chase* and take the final shot as precisely and accurately as possible to cause minimal suffering to the animal. Wild meat and fish are undoubtedly organic food without been treated with any kind of hormones or antibiotics, without any artificial dyes, and without being preserved for long periods of time with chemicals.

Animal well-fare should be everyone's concern, it is mistaken to assume that hunters are disrespectful and indifferent to animal welfare issues. The animal a hunter has taken never lived in cages, was never kept in small and crowded circumstances, and was never treated with antibiotics or with any other drugs. These animals experienced life – no matter how long or short – in a natural way. They lived naturally, reproduced naturally, gave birth naturally, fed naturally and behaved like free wild animals.

12. Conclusion

No matter how much it is talked about, or how many books are written, or how many talks are given, hunting remains to be the topic that splits the society in two. However, despite different views, the unconditional love and care for wildlife inevitably brings the two sides together, and that is the spark of hope for many.

Hunting has numerous positive effects on ecosystems when conducted properly. The main idea of sustainable harvest is to maintain animal populations at an economically and ecologically healthy level for the amount of space available. In this case local human communities benefit as well. It is very rare for a regulated wild animal population to cause any disturbances to rural communities. Trophy hunting is based on the idea of adaptive management which continuously monitors and evaluates the current circumstances and if necessary, revises management plan. Regulations like hunting seasons and quotas make sure that the populations are not harmed – making hunting rather safe management method for wildlife.

It is important to note that trophy hunting involves taking small numbers of individual animals and does not require highly developed infrastructure. The hunts are therefore high in value, but low in impact compared to other types of land uses like agriculture or tourism. Keeping populations fit is one of the priorities of sustainable hunting, therefore, removing weak and sick individuals is an important practice because it gives reproduction opportunity to stronger and younger individuals. Contributions that hunting makes to the economy are obvious, as many conservation programs have been financed by the revenues which were generated through hunting. These programs aim to recover most unique and endangered species populations, however, at the same time, some projects that are being financed with same revenue, aim to develop infrastructure in rural areas. With trophy hunting removed, countries risk to loose not only a major source of income and funding for conservation programs, but its valued wildlife altogether.

Wildlife management is a key part of biodiversity conservation. The unfortunate failure to understand the true essence of hunting is harming wildlife, which ultimately brings harm to humans as well. It is not the act of hunting but rather the purpose behind hunting that plays a key role for wildlife. Society calls for open-mindedness and encourages reflective thinking, yet it is this very society – our society – that rushes to conclusions. The negative attitude towards hunting not only effects the policy and workings of professionals but it impacts wildlife itself, the very thing society is trying so hard to protect.

References

Alley cat Allies, & . (2011). *The vacuum effect: why catch and kill doesn't work.*

Adler, P. S. (2014). Feral Cat Task Force: Findings and recommendations . the county of Kauai, Hawaii

Andersen, M., Martin, B., & Roemer, G. (2004). *Use of matrix population models to estimate the efficacy of euthanasia versus trap-neuter-return for management of free-roaming cats.* JAVMA.

Anthony Waldron, et al., "Targeting global conservation funding to limit immediate biodiversity declines," *Proceedings of the National Academy of Sciences of the United States of America* 110, no. 29 (2013)

Bhat Z.F, Bhat Hina, Pathak Vikas, Principles of Tissue Engineering, Academic Press, 2014.

Castillo, D., & Clarke, A. (2003). Trap/Neuter/Release Methods Ineffective in Controlling Domestic Cat "Colonies" on Public lands . *Natural Areas Journal* , 23:247-253.

Census of the number of hunters. FACE. 2010.

Chaber A.L., Allebone-Webb S., Lignereux Y., Cunningham A.A. & Rowcliffe J.M. (2010). – The scale of illegal meat importation from Africa to Europe via Paris. Conserv. Lett., 3, 317–323

Conover, Michael R. *Resolving Human-Wildlife Conflicts: The Science of Wildlife Damage Management*. CRC Press, 2001.

Conservation visions. 10 February 2017. Should we have trophy hunting in North America? With Shane Mahoney. https://www.youtube.com/watch?v=O6kEsuGx4D0&t=140s.

FACE. Middleton Angus. The Economics of Hunting in Europe: Towards a Conceptual Framework. 2014

Falk, H & Dürr, Salome & Hauser, Ruth & Wood, Kathy & Tenger, B & Lörtscher, M & Schüpbach-Regula, G. (2013). Illegal import of hushmeat and other meat products into Switzerland on commercial passenger flights. Revue scientifique et technique (International Office of Epizootics). 32. 727-39. 10.20506/rst.32.2.2221.

Fatjó, J., Bowen, J., García, E., Calvo, P., Rueda, S., Amblás, S., & Lalanza, J. F. (2015). Epidemiology of Dog and Cat Abandonment in Spain (2008–2013). *Animals* , 5(2), 426–441. .

Foley, P., Foley, J., Levy , J., & Paik, T. (2005). Analysis of the impact of trap-neuter-return programs on population of feral cats. JAVMA.

Fricke, Kent A.; Cover, Michael A.; Hygnstrom, Scott E.; Genoways, Hugh H.; Groepper, Scott R.; Hams, Kit; and VerCauteren, Kurt C., "Historic and Recent

Distributions of Elk in Nebraska" (2008). USDA National Wildlife Research Center - Staff Publications. 922

Gates, C.C., Freese, C.H., Gogan, P.J.P. and Kotzman, M. (eds. and comps.) (2010). American Bison: Status Survey and Conservation Guidelines 2010. Gland, Switzerland: IUCN.

Gunther, I., Finkler, H., & Terkel, J. (2011). *Demographic differences between urban feeding groups of neutered and sexually intact free-roaming cats following a trap-neuter-return procedure.* JAVMA.

Hamr, Josef, frank f. mallory, and Ivan filion. 2016. History of Elk (Cervus canadensis) restoration in Ontario. Canadian fieldnaturalist 130(2): 167–173.

Hervieux D, Hebblewhite M, Stepnisky M, Bacon M, Boutin S. Managing wolves (Canis lupus) to recover threatened woodland caribou (Rangifer tarandus caribou) in Alberta. 2014. NRC research press.

Huntley, B. J. (2001). South Africa's experience regarding alien species: impact and control. (O. Sandlund, Ed.) *pringer Science & Business Media*.

IUCN ESUSG (2006). Guidelines on Sustainable Hunting in Europe. IUCN-ESUSG WISPER.

IUCN SSC (2012). IUCN SSC Guiding principles on trophy hunting as a tool for creating conservation incentives. Ver. 1.0. IUCN, Gland.

Jędrzejewski W, Schmidt K, Theuerkauf J, Jędrzejewska B, Selva N, Zub K, Szymura L. 2002a. Kill rates and predation by wolves on ungulate populations in Białowieża Primeval Forest (Poland). Ecology 83: 1341-1356.

Jędrzejewska B., Jędrzejewski W. 2001. Ekologia zwierząt drapieżnych Puszczy Białowieskiej. Wydawnictwo Naukowe PWN, Warszawa.

Kurpiers, Laura & Schulte-Herbruggen, Bjorn & Ejotre, Imran & Reeder, DeeAnn. (2016). Bushmeat and Emerging Infectious Diseases: Lessons from Africa

Lerner, Jeff, et al. "What's in Noah's Wallet? Land Conservation Spending in the United States." *BioScience*, vol. 57, ser. 5, May 2017, pp. 419–423. 5, doi:10.1641/B570507.

Ljung, Per & Riley, Shawn & Heberlein, Thomas & Ericsson, Göran. (2012). Eat Prey and Love: Game Meat Consumption and Attitudes toward Hunting. Wildlife Society Bulletin. 36. 10.1002/wsb.208.

Lott, D. (2003). American bison. Berkeley, Calif,: university of California Press.

Mahoney , Shane, director. *Eating Eden to Extinction* . *YouTube*, Mystery Rnach, 22 Jan. 2019, www.youtube.com/watch?v=PpfSa4ToZ2o.

Mackintosh, Zig. *Elephant and the Pauper*. The Osprey Filming Company. 2015.

Marra, P. P., & Santella, C. (2016). Cat Wars: The Devastating Consequences of a Cuddly Killer. New Jersey : Princeton University Press .

Medina, F., Bonnaud, E., Vidal, E., Tershy, B., Zavaleta, E., Donlan, J., . . . Nogales, M. (2011). A global review of the impacts of invasive cats on island endangered vertebrates. *Global Change Biology*, 17, 3503–3510,.

Messmer A. Terry. (2000). The emergence of human-wildlife conflict management: turning challenges into opportunities. Elsevier Science.

Milton K, The Critical Role Played by Animal Source Foods in Human (*Homo*) Evolution, *The Journal of Nutrition*, Volume 133, Issue 11, November 2003, Pages 3886S–3892S, https://doi.org/10.1093/jn/133.11.3886S

Muller, L.I., R.J. Warren, and D.L. Evans. 1997. Theory and practice of immunocontraception in wild animals. Wildl. Soc. Bull. 25(2):504-515.

Neville, P. F. (1989). *Feral cats: management of urban populations and pest problems by neutering.* Chapman and Hall Ltd.

Nilsson M. L, Chapter 7 - Food, Nutrition, and Health in Sápmi, Nutritional and Health Aspects of Food in Nordic Countries,Academic Press,2018,Pages 179-195.

Nogales, M. A., Martin, B. R., Tershy, C. J., Donald, D., Veitch, N., & Puerta, B. J. (2004). A review of feral cat eradi- cation on islands. *conservation Biology*, 18:310–319.

Numbers in Kenya: What Are the Causes? PLoS ONE 11(9): e0163249. https://doi.org/10.1371/journal.pone.0163249

Ogutu JO, Piepho H-P, Said MY, Ojwang GO, Njino LW, Kifugo SC, et al. (2016) Extreme Wildlife Declines and Concurrent Increase in Livestock

Organ, J.F., V. Geist, S.P. Mahoney, S. Williams, P.R. Krausman, G.R. Batcheller, T.A. Decker, R. Carmichael, P. Nanjappa, R. Regan, R.A. Medellin, R. Cantu, R.E. McCabe, S. Craven, G.M. Vecellio, and D.J. Decker. 2012. The North American Model of Wildlife Conservation. The Wildlife Society Technical Review 12-04. The Wildlife Society, Bethesda, Maryland, USA

Orueta , J., & Ramos , Y. (2001). *Methods to control and eradicate non-native terrestrial vertebrate species.* Council of Europe Publishing.

Okarma H., Jędrzejewski W., Schmidt K., Kowalczyk R., Jędrzejewska B. 1997. Predation of Eurasian lynx on roe deer and red deer in Białowieża Primeval Forest, Poland. Acta Theriol. 42: 203-224.

Papadodimas, N. (2011) How do Hunting Organizations in Greece contribute in law enforcement mechanisms. European Conference: Illegal Killing of Birds Cyprus, Larnaka, 6 – 8 July 2011.

Public and Corporate Economic Consultants (PACEC) (2014). The Economic, Environmental and Social Contribution of Shooting Sport in UK. http://www.shootingfacts.co.uk/pdf/The-Value-of-Shooting-2014.pdf 12

Russell F, Zippin D, Fowler N. Effects of White-tailed Deer (Odocoileus virginianus) on Plants, Plant Populations and Communities: A Review. *American Midland Naturalist* [serial online]. July 2001.

Scallan, D. (2013) A Socioeconomic Assessment of Hunting in the Republic of Ireland. Report for the Federation of Field Sports of Ireland and the National Association of Regional Game Councils. February 2013

Selva N., Jędrzejewska B., Jędrzejewski W., Wajrak A. 2005. Factors affecting carcass use by a guild of scavengers in European temperate woodland. Can. J. Zool. 83: 1590-1601

Stavisky . (2012). Demographics and economic burden of un-owned cats and dogs in the UK: results of a 2010 census. *BMC Veterinary Research*, 8:163.

Tasker, L. (2007). *Stray Animal Control Practices (Europe).*

The Northeast Furbearer Resources Technical Committee. *Trapping and Furbearer Management in North American Wildlife Conservation.* 2015, www.dec.ny.gov/docs/wildlife_pdf/trapfurmgmt.pdf.

U.S. Department of the Interior, U.S. Fish and Wildlife Service, and U.S. Department of Commerce, U.S. Census Bureau. 2016 National Survey of Fishing, Hunting, and Wildlife-Associated Recreation.

U.S. Fish & Wildlife Service. 2016 National Survey of Fishing, Hunting, and Wildlife-Associated Recreation, National Overview. August 2017

Vernon R. Booth (2010): The Contribution of Hunting Tourism: How Significant is this to National Economies? In Contribution of Wildlife to National Economies. Joint publication of FAO and CIC. Budapest.

Williamson D, Bakker L.The Bushmeat Crisis in West-Africa
An indicative overview of the situation and perception. FAO.

Wyler L, Sheikh P (2008) International illegal trade in wildlife: threats and U.S. policy. CRS Report for Congress, March 3, 2008, 49 pp

Made in the USA
Las Vegas, NV
10 April 2022

47225149R00039